IMAGES
of Aviation

FLIGHT RESEARCH AT
NASA LANGLEY
RESEARCH CENTER

NASA Langley Research Center, Hampton, Virginia. NASA 515, a Boeing 737 research aircraft, flies over the NASA Langley Research Center, the first and oldest civilian-run government aerospace research center in the United States. The center currently employs 3,300 civil servants and contractors. Langley is internationally recognized as a home of innovation and cutting-edge achievement in aeronautics and space technology, helping to make America a world leader in aerospace. (NASA.)

On the Cover: Langley test pilot John P. "Jack" Reeder poses with a NACA/Air Force McDonnell F-101A Voodoo jet fighter at Langley Field in 1956. The aircraft was flown by NACA Langley in sonic boom studies in addition to handling and performance evaluations. (NASA.)

IMAGES
of Aviation

FLIGHT RESEARCH AT NASA LANGLEY RESEARCH CENTER

Mark A. Chambers

ARCADIA
PUBLISHING

Published by Arcadia Publishing
Charleston SC, Chicago IL, Portsmouth NH, San Francisco CA

Printed in the United States of America

Library of Congress Catalog Card Number: 2006940726

For all general information contact Arcadia Publishing at:
Telephone 843-853-2070
Fax 843-853-0044
E-mail sales@arcadiapublishing.com
For customer service and orders:
Toll-Free 1-888-313-2665

Visit us on the Internet at www.arcadiapublishing.com

This pictorial history is a tribute to the personnel of the Flight Research Division of NACA/NASA Langley, who throughout the years worked with unwavering dedication toward the advancement of flight.

CONTENTS

ACKNOWLEDGMENTS

The author would like to thank several individuals who helped to make this pictorial history a reality. Keith Loftin, son of the late Laurence K. "Larry" Loftin Jr., former director for aeronautics at the NASA Langley Research Center, contributed his father's extensive photograph collection for this effort. Former Langley test pilots including the late John P. "Jack" Reeder and the late Robert A. Champine also contributed heavily to this publication by making their personal photograph collections available to the author.

Current and former staff members of Langley were especially helpful in the preparation of this publication. The author thanks A. G. "Gary" Price, retired head of NASA Langley Research Center Office of External Affairs, for providing access to special photographic projects undertaken by the Langley Office of Public Affairs. Special thanks also go to the staff of the Langley Photographic Laboratory for assisting in photographic reproduction, Patricia A. West of Langley for data and administrative assistance, and Barbara Trippe of the Langley Flight Operations Office for providing access to historical photographs and official pilot information. Garland Gouger and the staff of the Langley Technical Library provided invaluable assistance in locating reports and other documents. Finally, the author would like to thank his father, Langley retiree Joseph R. Chambers, for serving as technical consultant.

Special thanks also go to Courtney Hutton, Kate Crawford, and the staff of Arcadia Publishing for their assistance and dedicated support for this project.

It should be noted that this publication is not an official NASA publication and that the opinions expressed herein and any content errors are the author's responsibility.

INTRODUCTION

The Atlantic Coastal Region of the United States is known throughout the aviation world for its significance in American flight research activity. Notable flight research achievements in this region include: successful flights of Samuel Pierpont Langley's unmanned powered flying model (Aerodrome No. 5) on May 6, 1896, from a houseboat on the Potomac River; the Wright brothers' successful manned, powered flights in their 1903 Flyer on the dunes of Kitty Hawk, North Carolina, on December 17, 1903; the Wrights' successful demonstration of their "Military Flyer" for the U.S. Army at Fort Myer, Virginia, on July 2, 1909; Eugene Ely's successful takeoff in a Curtiss Pusher from a makeshift wooden deck on the bow of the armored cruiser USS *Birmingham*, anchored in the waters of Hampton Roads, Virginia, on November 14, 1910; and the aerial exploits of legendary figures in the field of aviation, such as Billy Mitchell and Eddie Stinson, at the Atlantic Coast Aeronautical Station established by aviation pioneer Glenn Curtiss in Newport News, Virginia, in 1915.

On March 3, 1915, the U.S. Advisory Committee for Aeronautics was established by a congressional rider to the Naval Appropriations Act, "to supervise and direct the scientific study of the problems of flight, with a view of their practical solution." On April 23, 1915, the word "National" was included during the initial meeting of the committee, and it was henceforth known as the National Advisory Committee for Aeronautics (NACA). On November 22, 1916, the NACA Executive Committee gave formal approval for a suggestion from the NACA Sub-Committee that the location now known as Langley Field in Hampton, Virginia, be designated for NACA and U.S. Army Air Service use for national defense and airplane flight testing, research, and development. Hampton was deemed an ideal location for such activities because of the area's relatively mild climate and hospitable weather conditions, flat land and coastal waters for seaplane/flying boat operations, and proximity to Washington, D.C. The site was also near ship manufacturing assets in Newport News, Portsmouth, and Norfolk.

On July 17, 1917, a ground breaking was held for the Langley Memorial Aeronautical Laboratory (LMAL) at Langley Field. The site included 1,650 acres and was procured for a sum of $290,000. With operations being controlled by the NACA, the LMAL consisted of five divisional elements: the Aerodynamics Division, Power Plants Division, Technical Service Division, Flight Operations Division, and Property and Clerical Division. Leigh M. Griffith was chosen to serve as the laboratory's first engineer in charge, or director. Some of the lab's first facilities included a research laboratory administration building, a technical library, photographic laboratory, and headquarters for the Aerodynamics, Power Plants, Technical Service, and Flight Operations Divisions. In the immediate years to follow, two aerodynamic laboratories, two engine dynamometer laboratories, an airplane hangar complete with repair sections for accommodating aircraft employed in flight research operations, a service building complete with an instrument laboratory, a drafting room, and machine and woodworking shops became augmentations to the LMAL. In the years during and immediately after World War I, the mission of the LMAL was clear: to find solutions to aeronautical problems, conduct research on fundamental phenomena, and keep pace with aeronautical research and development in Europe.

In 1918, work commenced on an important tool designed to augment flight research studies at the LMAL, the Five Foot Atmospheric Wind Tunnel (AWT). With its completion in 1919, engineers attempted to simulate flight conditions on the ground and correlate data with actual flight tests. In June 1919, the NACA's first two research aircraft, two Curtiss JN-4H Jenny aircraft procured from the U.S. Army Air Service, arrived at Langley for flight testing. Thomas Carroll and Edmund T. "Eddie" Allen were hired by the NACA to serve as the agency's first two test pilots. Their first mission was to correlate flight data with data gathered through wind-tunnel studies of Jenny models. The pilots also flew their aircraft in stability and control studies. The agency's first flight tests revealed that although the airplanes were similar in appearance, they possessed many differences in structural details that influenced their flying characteristics. One of the aircraft's propellers was warped and its wings manufactured incorrectly. Following the first test, the research aircraft were outfitted with special instruments for further fight testing. These instruments included an altimeter, tachometer, and airspeed indicator.

The formal technical report generated from the first flight test directed attention toward the need for a special research pilot, one with extensive experience in the field of aeronautical engineering. Over the years, NACA/NASA Langley would employ numerous dedicated pilots that fit this description to a T.

From these meager beginnings, the modest flight research program at Langley quickly burgeoned into a full-fledged, bona fide flight research operation with a multitude of flight test subjects and projects. Throughout the 20th century, the flight research operation at Langley helped to make innovative dreams and perfection in aeronautics a reality. The laboratory helped to spur progress in American aviation during the golden age of aviation in the 1920s and 1930s. Flight research at Langley helped the United States, as well as its Allies, win World War II by verifying, through flight tests, the results of modifications to warplanes initiated through extensive wind-tunnel tests carried out at the request of the military services. Following World War II, Langley's Flight Research Division helped the nation's military services successfully make the transition into the jet age and helped keep the country at the forefront of aeronautical technology, ahead of the Soviet Union and the Warsaw Pact during the cold war. Immediately following World War II, Langley personnel involved in flight research made the trek to Muroc, California, to take part in pioneering supersonic and high-speed aircraft flight research. NACA involvement in these studies later evolved into the Dryden Flight Research Center, today's premier flight research center for NASA.

In addition, Langley's Flight Research Division helped provide technology used by designers of large commercial jet transports for domestic and international travel. Benefits derived from both flight and wind-tunnel research of propeller-driven aircraft carried out at the laboratory during World War II and immediately following the war were later applied to civilian general aviation or light aircraft used as personal transports. Flight research performed by Langley in this area helped to make the general aviation market in the United States during the 1970s boom.

In October 1958, the National Advisory Committee for Aeronautics, under authorization from President Eisenhower, officially became incorporated into a new agency known as the National Aeronautics and Space Administration (NASA). This transformation and new emphasis on space flight had a dramatic impact on Langley's flight research operations. The transition, which included a name change from the Langley Aeronautical Laboratory (LAL) to the Langley Research Center (LaRC), was marked by a transferal of fixed-wing, high-performance flight-test research projects from Langley to NASA Dryden Flight Research Center in the Mojave Desert near Edwards Air Force Base in California and the transfer of a substantial number of Langley flight research personnel to space activities at the new NASA Manned Spacecraft Center (now Johnson Space Center) in Houston, Texas, in the early 1960s. During that time, the emphasis of Langley's flight research efforts shifted to Vertical Take Off and Landing (VTOL) projects and rotorcraft projects. Langley and its cast of piloting experts became a pioneering force in these areas, and the research center's exploits in these arenas had an international reach.

The 1970s, 1980s, and 1990s witnessed the use of unique "flying laboratory" aircraft at Langley that helped to pioneer high-tech cockpit systems that eased piloting workload and enhanced operational safety; to support studies of environmental factors as meteorological and atmospheric phenomena affecting aircraft operations and life on Earth; and to study phenomena induced by aircraft operations such as contrails, sonic boom, and wake vortex formation. These programs were interspersed with advanced general aviation studies intended to flight test new state-of-the-art technologies designed to enhance the country's small aircraft transportation system and revitalize the general aviation industry. The contributions of the center's storied past in flight research projects to the advancement of American aviation have not been forgotten. This pictorial survey serves as a tribute to the rich legacy of flight research contributions by those who served with dedication in the Flight Research Division of the NASA Langley Research Center.

One

The Formative Years
1915–1922

Following the creation of the National Advisory Committee for Aeronautics (NACA) in 1915, the agency pursued the construction and development of the nation's first civilian-led aeronautics research laboratory, the Langley Memorial Aeronautical Laboratory (LMAL) in Hampton, Virginia.

At the LMAL, the agency conducted its first flight tests with Curtiss JN-4H Jenny trainers acquired from the Army Air Service in 1919 and began correlating flight data with data generated from wind-tunnel studies conducted in the laboratory's Five Foot Atmospheric Wind Tunnel. The agency's first two test pilots, Edmund T. "Eddie" Allen and Thomas Carroll, performed numerous test flights in the laboratory's two Jennies, conducting pressure distribution studies over the wings and structures of the aircraft as well as basic fundamental aerodynamic flight studies. These research studies were so thorough that the pilots spotted geometric inaccuracies, including a warped propeller and incorrectly manufactured wings. Such were the problems encountered in early aircraft manufacturing.

During the early 1920s, Langley test pilots documented the basic handling and takeoff and landing qualities of military aircraft after performing flight tests in World War I aircraft. The LMAL also acquired a Vought VE-7 trainer from the navy for flight tests. Along with the lab's fleet of Jenny trainers, the VE-7 was used to perform flight tests aimed at the development of navigational aids and understanding flight qualities in adverse weather.

The foundation for a rich tradition in flight research had been forged.

NACA's First Flight Research Aircraft and Project. The NACA's first flight research aircraft, a JN-4H Jenny on loan from the army, is being prepared for takeoff on a research mission (above). Such flights were performed by the agency's first two test pilots, Edmund T. "Eddie" Allen and Thomas Carroll. Interestingly, these two pioneers of flight research had no formal education in aeronautical engineering. The initial NACA test flights in the JN-4H were intended to define the basic flight characteristics of aircraft. Below, one of NACA Langley's two Jenny aircraft trails a pitot-static tube to measure airspeed in one of the NACA's first flight research missions. (NASA.)

WORLD WAR I FOES. In this snow-covered scene, NACA Langley pilots Eddie Allen (left) and Tom Carroll (right) pose in front of a trio of World War I combatants: from left to right, a French-built Spad XIII, British-built SE5a, and German Fokker D-VII. The aircraft were used by Langley to study the handling and flight qualities of military aircraft during the early 1920s. Allen later left the NACA to work for the Boeing Aircraft Company in Seattle, Washington. Tragically, Allen lost his life while test flying a Boeing B-29 Superfortress four-engine bomber prototype, designated the XB-29, which crashed near Boeing Field on February 18, 1943. (NASA.)

FIRST NACA LANGLEY HANGARS. This photograph, taken in 1931, shows the first hangars at NACA Langley with the lab's Fairchild transport airplane visible at the right. Also visible in the photograph is a specially rigged Ford Model A used to crank-start the engines. (NASA.)

EARLY LINEUP OF FLIGHT RESEARCH SUBJECTS AT LANGLEY. An early lineup of JN-4H Jennies and a single Vought VE-7 (right) are readied for flight research missions. In addition to other uses, the aircraft were used to perform flight tests aimed at the development of navigational aids and understanding flight qualities in adverse weather. (NASA via Jack Reeder.)

EARLY LANGLEY FLIGHT RESEARCH PIONEERS. Aeronautical engineer Henry J. E. Reid (front) and Tom Carroll (rear) prepare for a research mission in the LMAL's Vought VE-7. Reid later became the director of the LMAL at the time of some of the lab's greatest achievements. (NASA via Laurence K. Loftin.)

Two

THE GOLDEN AGE OF AVIATION
1923–1941

During the 1920s and 1930s, aircraft designs grew more complex. Fabric and wooden aircraft structures gave way to sturdier aluminum aircraft structures. Langley researchers focused on structures, aerodynamics, and developing innovations designed to enhance flight. In 1923, the LMAL conducted some of the first American in-flight airflow visualization studies.

During this time period, Langley also began studying seaplane performance. By the early 1930s, Langley supplemented its flight studies with ground studies of flying boat and seaplane hulls and pontoons that were drawn through a long tow tank, Tow Tank No. 1. This led to dramatic performance improvements in seaplanes and flying boats. Tow Tank No. 1 was designated a National Historic Landmark in 1985.

While developing land-based aircraft during the mid- to late 1920s, aircraft manufacturers became concerned with performance degradation caused by drag created by airflow over the cylinders and engine stacks of propeller-driven aircraft. Langley began experimenting with various types of engine cowlings capable of minimizing drag while permitting adequate cooling for the engine. These experiments resulted in the derivation of the famous NACA cowling by engineer Fred Weick and his staff. With this ingenious innovation, later adopted by most airplane designers throughout the world, the LMAL won the coveted Collier Trophy for 1929.

Throughout the mid- to late 1920s and 1930s, Langley conducted numerous flight experiments with a variety of aircraft wing configurations. New flaps and wing leading-edge devices that generated greater lift were also tested and proven. Langley performed some of the first test flights in American-built autogiros. These studies yielded some of the first formal documentation on these precursors to the modern-day helicopter.

Langley broke new ground in aircraft icing and deicing research. A unique deicing apparatus was developed and flight tested by the laboratory. During the 1930s, Langley conducted precedent-setting stall alleviation and spin-recovery studies in a variety of aircraft. Langley explored the full potential of high-altitude flight using a British invention, the supercharger.

By 1941, the LMAL was conducting some of the first extensive flight studies in severe thunderstorms to investigate problems associated with wind gusts and improving operations in adverse weather. The lab's pilots also helped develop requirements for airplane flying qualities, a forerunner to today's military specifications.

AIRFLOW VISUALIZATION USING SMOKE. In 1923, NACA Langley conducted one of the first in-flight flow visualization studies when smoke was trailed over the wings and control surfaces of a Martin MO-1, on loan from the Naval Air Station (NAS) Hampton Roads (now NAS Norfolk). The MO-1 was the U.S. Navy's first monoplane aircraft and made use of a thick metal wing for reinforced structural integrity, but it possessed several nagging deficiencies. During the flight investigations, the MO-1 was piloted by NACA test pilot Tom Carroll, while NACA Langley engineer David L. Bacon flew in the observer's position located in the midsection of the fuselage, photographing the smoke as it flowed over the wing. (NASA.)

SIDE VIEW OF MARTIN MO-1. This is a side view of the Martin MO-1 aircraft in storage at the NACA Langley flight hangar. One suggestion offered by the NACA to the navy and the aircraft manufacturer, Martin, to improve the airplane's performance was to reduce the size of a large gap between the wing and aileron. (NASA.)

VENERABLE VIKING. This British-built Vickers Viking IV amphibian was originally obtained by the U.S. Navy in 1921 and loaned to the NACA in 1924 so that the agency could acquire practical experience in operating and studying the design of amphibians. In addition to undergoing performance and handling studies, the aircraft was flight tested with various types of propellers. (NASA.)

SIXTH ANNIVERSARY DISPLAY OF FLIGHT AT NACA LANGLEY, 1925. NACA flight division personnel pose for a sixth-anniversary photograph of flight at Langley in 1925. The aircraft is a Vought VE-4. Langley pilots Tom Carroll (standing second from left), Paul King (seated far left), and Bill McAvoy (seated third from right) can be seen in the picture. (NASA via Jack Reeder.)

Sperry M-1 Messenger. An interesting series of flight tests at Langley involved the use of an Army Sperry M-1 Messenger, which was flown with various airfoils between 1925 and 1927, most notably the variable-camber wing. Correlations between wind-tunnel studies and flight investigations were made in a relatively short period of time. Advancements in flight research technology, such as the recording devices used on the Messenger, were making flight research a more exact science. (NASA.)

Wright WF3W-1 Apache Seaplane. Outfitted with floats, a Wright WF3W-1 Apache is revved up for a flight research mission on the Back River, located adjacent to Langley Field in 1927. The NACA flight tested the Apache Seaplane in studies aimed at examining the engine performance of the aircraft as well as suitable cowling shapes for its engine. A supercharged land version of the Apache was later flown to a record-setting altitude by navy lieutenant Apollo Soucek in 1929 at NAS Anacostia, D.C. Soucek stunned aviation authorities around the world by reaching an altitude of 40,366 feet in the airplane. (NASA.)

SUPERCHARGED STALWART. The British-built DeHavilland DH-4B served as a stalwart of early American military aviation, serving in both the Army Air Corps and the Marine Corps. The NACA used this supercharger-equipped DH-4B in flight research missions aimed at unlocking the secrets of high-altitude flight. The DH-4B provided NACA officials the opportunity to test the revolutionary supercharger, an English invention, on an air-cooled American-built Liberty engine. (NASA via Laurence K. Loftin.)

PIONEER OF THE HEAVENS. Langley test pilot Bill McAvoy dons a specially designed high-altitude flight suit while posing for a publicity photograph next to the NACA supercharged DeHavilland DH-4B. (NASA via Jack Reeder.)

HIGH-FLYING APACHE. A Langley test pilot, sporting high-altitude flight gear, poses beside a Wright Apache land-based aircraft, acquired by the NACA for high-altitude flight research studies, prior to a mission in 1928. (NASA.)

CURTISS TS-1 ON FLOATS. This Curtiss TS-1 seaplane, produced by the Naval Aircraft Factory in Pennsylvania, featured an air-cooled radial engine and was flight tested by NACA Langley primarily in studies to determine the pressure distribution across the floats. Performance and handling studies of the seaplane were also conducted by Langley. (NASA.)

HIGHLY MANEUVERABLE BIPLANE. This Consolidated NY-2, loaned to NACA Langley by the navy, was one of the first aircraft to be equipped with NACA leading-edge slats for enhanced maneuverability and stability in 1928. (NASA.)

FIRST NACA COWLING SUBJECT. This army Curtiss AT-5A became the first aircraft to utilize the revolutionary NACA cowling developed by Langley engineer Fred Weick and his staff. When the cowling was applied to the AT-5A in 1928, the aircraft's speed and performance increased dramatically from 118 to 137 miles per hour without any modifications to the engine. (NASA.)

Manufacturing Cowling Shapes. NACA mechanics and shop workers are at work developing various aircraft cowling shapes around the late 1920s. (NASA.)

Installing a NACA Cowling. The NACA cowling was also applied to civil aircraft. Here a team of NACA mechanics installs a revolutionary NACA cowling on a Fokker Trimotor (Atlantic C-2A, aircraft 28-1230) commercial airliner. (NASA.)

LANGLEY FLIGHT OPERATIONS GROUP IN FRONT OF FOKKER TRIMOTOR. In this 1928 photograph, the Langley Flight Operations Group poses in front of a NACA cowling–outfitted Fokker Trimotor (Atlantic C-2A, aircraft 28-123). Three pilots who performed most of the test piloting during the early days of flight research at the laboratory can be seen in the picture: Bill McAvoy (standing third from left), Tom Carroll (standing fourth from left), and Melvin Gough (seated fifth from left). (NASA via Jack Reeder.)

STRUCTURAL TEST BED. This army Boeing PW-9 pursuit biplane was used by NACA Langley as a structural test bed to experiment with new materials used to cover the airframe. Mechanics at Langley later installed a metal skin (aluminum) on the fuselage in 1928 among other elements of the airplane. (NASA via Laurence K. Loftin.)

Boeing PW-9 Pursuit Biplane. This army PW-9 pursuit biplane was used by NACA Langley for structural testing in 1928. The aircraft has been outfitted with a metal-covered fuselage. (NASA.)

Spin Test Bed. A navy Consolidated NY-1 was used by NACA Langley in extensive spin tests in 1929. Flight tests of this airplane were supplemented with wind-tunnel spin tests in the Langley Spin Tunnel. (NASA.)

BIG CURTISS FLYING BOAT. This huge Curtiss H-16 Flying Boat, another Naval Aircraft Factory product, was used by NACA Langley in hull pressure distribution investigations in 1929. The aircraft is resting on a dolly on the seaplane ramp at Langley located near the Back River. (NASA.)

VARIABLE ANGLE COWL. One of many cowling shapes the NACA studied was the variable angle cowl, conceived by engineers at Langley. Here the cowling was applied to a navy Curtiss XF7C-1 Seahawk, a carrier-based fighter, in 1930. (NASA.)

PITCAIRN PCA-2 AUTOGIRO. In July 1931, Langley test pilots Bill McAvoy and Melvin Gough conducted a series of test flights in America's first autogiro, the Pitcairn PCA-2. The flight tests were the first to yield quantitative data regarding the handling qualities of this forerunner of the modern-day helicopter. (NASA.)

SUBMARINE FLYING BOAT. Langley test pilot Melvin Gough poses beside a Loening XSL-1 flying boat at NAS Hampton Roads (now NAS Norfolk) in 1931. The aircraft underwent performance and handling tests at Langley. Gough was the first NACA test pilot with a degree in aeronautical engineering, obtained from Johns Hopkins University, and he later rose in the Langley ranks to the position of chief test pilot. In 1958, Gough became director of NASA activities at the Atlantic Missile Range, Cape Canaveral, Florida. (NASA.)

SWIFT SEVERSKY. The floats on this Seversky Amphibian were the first to be tested in Langley Tow Tank No. 1, which consisted of a huge channel filled with water in which seaplane floats, flying boat hulls, and pontoons were drawn through the water for testing. The tests resulted in significant design improvements that led to the aircraft setting a new international speed record during the 1930s. (NASA.)

TOW TANK NO. 1. A flying boat model is prepared for tests in Tow Tank No. 1 in 1945. (NASA.)

EAST AREA HANGAR CONSTRUCTION, NOVEMBER 6, 1931. Construction progresses on a new aircraft hangar in the old East Area of NACA Langley. A new hangar was needed to store and house the growing number of research aircraft flown by the NACA. (NASA.)

EAST AREA HANGAR COMPLETED, 1932. Shown is the completed aircraft hangar in 1932. Note the Pitcairn PCA-2 Autogiro at the hangar entrance. (NASA.)

A STEP TOWARD MODERNIZATION. The Grumman XFF-1 represented a step toward modernization in fighter design. The aircraft was the first naval fighter to feature an enclosed cockpit with canopy and retractable landing gear. Langley test pilot Bill McAvoy performed the majority of test flights in this aircraft beginning in 1932. (NASA.)

HALL ALUMINUM XP2H-1. In 1932, Langley test pilot Bill McAvoy and a team of engineers were detached to the Norfolk Naval Base to perform flight evaluations of this huge four-engine biplane flying boat that featured an aluminum hull and structural components. Engineers also performed pressure distribution tests on the hull. (NASA.)

HIGH-LIFT FAIRCHILD F-22. This Fairchild F-22 was equipped with a high-lift leading edge device that permitted greater lift for the parasol wing design. The NACA used this aircraft in high-lift flight experiments in 1932. (NASA.)

FOWLER-FLAP FAIRCHILD F-22. The Fairchild F-22 was later equipped with a Fowler flap for low-speed performance testing in 1932. The Fowler flap significantly reduced landing speeds. (NASA.)

GULL-WINGED PARASOL DESIGN. The Douglas YO-31A featured a parasol wing design that was evaluated both in the Langley Full Scale Wind Tunnel and in flight tests in 1932. Here the aircraft is being tested in the huge Full Scale Wind Tunnel, where actual aircraft could undergo aerodynamic testing. (NASA.)

BOEING F4B-2. In further tests of aircraft spin and recovery capabilities, the NACA equipped this F4B-2 with a spin-recovery parachute on the tail of the aircraft. The parachute concept was first used by the British on a Sopwith Camel that was released from a dirigible in England and flown into a spin. Extensive spin tests were conducted with the F4B-2 at Langley throughout the early 1930s. (NASA.)

CLOSE-UP OF THE F4B-2 SPIN-RECOVERY PARACHUTE. In this photograph, the spin-recovery chute can be seen folded on a platform above the vertical tail of the aircraft. (NASA.)

BOEING F4B-2 WITH T-TAIL. The Boeing F4B-2 was also later outfitted with a T-tail for improved control, stability, and enhanced spin-recovery characteristics. (NASA.)

BIRTH OF THE TRICYCLE LANDING GEAR. In 1933, Langley engineer Fred E. Weick, along with a team of colleagues, designed and developed a home-built aircraft known as the W-1A. It featured a pusher design complete with tricycle landing gear. NACA Langley conducted flight tests of the airplane, and the tricycle landing gear was later incorporated in several propeller-driven aircraft designs and virtually all modern-day jet aircraft designs. The W-1A was later evaluated in the Langley Full Scale Wind Tunnel, as seen here in 1934. (NASA.)

BOEING P-26 PEASHOOTER. This Boeing P-26 Peashooter, based at Bolling Field near NAS Anacostia, D.C., was flight evaluated in 1934 for modifications and improvements in the design that came about as a result of Full Scale Wind Tunnel tests. The Peashooter was the army's first all-metal monoplane pursuit fighter and served in fighter or pursuit squadrons until the latter 1930s. (NASA via Larry Loftin.)

KELLETT YG-1 AUTOGIRO. Engineer John B. Wheatley poses beside a Kellett YG-1 autogiro evaluated by the NACA in 1936. Wheatley wrote and compiled the majority of the NACA's research documents regarding autogiro research. In addition to being flight tested, this YG-1 underwent aerodynamic studies in the Langley Full Scale Wind Tunnel. (NASA.)

SLOTTED-COWL GOSHAWK. This navy Curtiss XBFC-1 (XF11C-1) Goshawk was fitted with a special slotted cowl and studied in flight experiments by NACA Langley, which also included pilot visibility studies, in 1937. The aircraft was designed to serve as a fighter-bomber. (NASA.)

SEVERSKY P-35 FIGHTER. This army Seversky P-35 fighter was flown by NACA Langley in handling and performance studies, as well as engine cowling studies in 1937. The aircraft design incorporated standard pre–World War II fighter design features and was a precursor to some of the most famous American World War II fighter designs. (NASA.)

A WEIGHTY ISSUE. A navy Grumman F3F-2 pursuit biplane undergoes weight, balance, calibration, and landing gear testing in the flight hangar at NACA Langley in 1937. The F3F-2 was an advanced version of Grumman's earlier XFF-1 and also possessed retractable landing gear. (NASA.)

CHUTE SPIN-RECOVERY SYSTEM. A close-up shows the parachute spin-recovery system on the rear of a Grumman F3F-2 pursuit biplane at NACA Langley in 1938. This aircraft was extensively test flown by Langley pilots Mel Gough and Bill McAvoy. (NASA.)

CONSOLIDATED XPB2Y-3 FLYING BOAT. In 1938, Langley pilots and engineers were detached to NAS Norfolk, Virginia, to conduct test flights in the Consolidated XPB2Y-3 Flying Boat prototype. Pressure distributions across the hull were measured and the information supplied to the navy in addition to an analysis of the flight performance and handling characteristics of the aircraft. (NASA.)

PRE–WORLD WAR II FLYING FORTRESS. This pre–World War II Boeing B-17 Flying Fortress was flown by NACA Langley pilots in 1938 in studies of the aircraft's wing structure in addition to performance and handling qualities. (NASA via Laurence K. Loftin.)

BIG BOMBER. The army Boeing XB-15 long-range bomber was test flown at NACA Langley in investigations of gust loads on structural elements of the airplane in 1938. The airplane was never mass-produced but established a world nonstop distance record in 1939 when it was flown from Langley Field, Virginia, to Chile to transport medical supplies to the then–earthquake ravaged nation. The airplane was in the air for almost 30 hours. (NASA.)

BIG SIKORSKY FLYING BOAT. In 1939, NACA Langley personnel visited NAS Norfolk to conduct flight studies of the navy Sikorsky XPBS-1 flying boat. The XPBS-1 was one of many exotic flying boats designed by the Sikorsky Company under the leadership of rotorcraft pioneer Igor Sikorsky, a Ukranian immigrant who came to the United States in 1919. The XPBS-1 was not mass-produced. Among the studies on the aircraft performed by the NACA were hull pressure distribution and performance and handling qualities testing. (NASA.)

PERFORMING PRESSURE DISTRIBUTION TESTS ON THE HULL OF THE SIKORSKY XPBS-1. This close-up view of the Sikorsky XPBS-1 at NAS Norfolk in 1939 shows pressure instruments strategically placed by NACA Langley engineers on the hull of the aircraft for testing. (NASA via Laurence K. Loftin.)

DEICING SOLUTION. One of the ageless problems encountered in flight has been the problem of ice buildup on the wings and critical flight structures of airplanes in adverse weather. In 1939, NACA Langley engineers developed a special airfoil test section that was heated by a series of coiled heating elements. This special airfoil deicing apparatus was test flown from 1939 to 1940 by Langley pilots on a Martin XBM biplane. (NASA.)

GYROPLANE ON FLOATS. In 1939, NACA Langley conducted flight experiments with one of the first hingeless rotor rotorcraft, the Wilford XOZ-1 Gyroplane. Engineers specifically studied the aerodynamic qualities and performance of the blades on the rotorcraft and favored the versatility of the hingeless rotor design as opposed to other rotor designs. Here the Gyroplane, which served with the navy, is being prepped for a research flight on the Back River. (NASA via Jack Reeder.)

FAIRCHILD XR2L-1(22). NACA Langley used this Fairchild XR2L-1(22) to flight test special flaps designed to improve landing performance in 1940. (NASA.)

LANGLEY'S FIRST "STORM-CHASER" AIRCRAFT. NACA Langley used this army Lockheed XC-35, the first aircraft in the world to have a pressurized cabin, as its first "storm-chaser" aircraft in 1940. The aircraft was flown into severe thunderstorms to investigate flight in wind gusts, heavy rain, and lightning. In addition to storm research, the aircraft was used to investigate contrail formation and in high-altitude studies. The aircraft is seen here configured with a rain collector boom attached to the nose for storm research. (NASA via Robert Champine.)

Three

SOWING THE SEEDS OF VICTORY
1942–1945

With the United States' entry into World War II in late 1941, aircraft development efforts were doubled. This meant a drastic increase in work for personnel at NACA Langley. Aerodynamically improved designs emerging from tests in Langley's Full Scale Wind Tunnel needed to be flight proven to confirm the soundness of potential design improvements.

In addition to verifying drag reduction and inlet recovery efforts, the Langley Flight Research Division tested propeller efficiency and longitudinal stability. Roll performance of fighter aircraft and all-moving horizontal stabilizers were also studied. The dramatic increase of work led to an updating of military handling requirements by the NACA.

During the war, Langley pilots conducted evaluations of foreign aircraft. These evaluations provided valuable information regarding the performance of both Allied and Axis aircraft. Flight research of advanced airfoils, such as the NACA laminar air foil, was also carried out. Langley pilots performed flight evaluations of the first practical helicopter, the Sikorsky HNS-1 Hoverfly.

One of the more interesting studies conducted by NACA Langley and the army involved ditching a Consolidated B-24D Liberator in the James River. These studies benefited aircrews of heavy bombers operating over vast areas of ocean in both the European and Pacific theaters.

As the war progressed, military pilots began to push the limits of their piston-engine fighter aircraft. Pilots began to experience compressibility problems with several fighter types, including the Lockheed P-38 Lightning. Engineers and pilots alike developed an interest in conducting tests in a speed range that exceeded the limits of these fighter aircraft, known as the transonic flight regime (speeds approaching and passing through the sound barrier).

Scientists and engineers at Langley developed two methods for studying transonic phenomena. One method known as the wing flow method, conceived by Dr. Robert Gilruth, involved installing various types of wing models on a special wing test section of a P-51D Mustang and forcing the aircraft into a steep dive. The other method involved dropping bomb drop models from underwing pylons on bomber aircraft. The bomb drop models reached transonic speeds as they fell and were tracked via telemetering.

Langley pilots were required to become proficient and master flight in several types of fighters, a dive bomber, a torpedo bomber, and a heavy bomber, in addition to making routine flights in civilian training aircraft.

WEED-OUT TRAINER. Langley used this Ryan ST trainer during World War II to screen test pilot candidates during interviews for research pilot jobs at the laboratory. The chief test pilot, Mel Gough, would take the candidate up in the aircraft and put the airplane through elaborate flight maneuvers to test the poise and stamina of the prospective test pilot candidates. The aircraft was also used for basic civilian aircraft pilot proficiency flights. (NASA via Laurence K. Loftin.)

AERODYNAMIC DRAG. Drag reduction was the emphasis of flight research on this Curtiss P-40B Tomahawk immediately prior to and during the early portion of World War II. The fairings under the wing used to house the landing gear were modified and improved in design among other changes to permit improved aerodynamic performance. This and other design modifications were later implemented in the army's Tomahawk fleet. (NASA.)

BATTLE OF BRITAIN VETERAN. The Hawker Hurricane Mk. IIA, which excelled as an interceptor during the Battle of Britain, was evaluated in late 1941. The Hurricane was part of NACA flight research studies designed to eliminate drag and stall problems created by the British practice of placing canvas-type fabric over the gun ports of the aircraft. As a result of the NACA flight studies, the British were able to prevent further incidents from occurring in their Hurricanes. (NASA.)

ANOTHER BRITISH VETERAN. The Supermarine Spitfire Mk. II, which also excelled during the Battle of Britain, was evaluated by Langley in early 1942. The Spitfire studies also included gun port covering effects as well as maneuverability and handling quality evaluations. (NASA via Laurence K. Loftin.)

NAVY WILDCAT. The navy Grumman F4F-3 Wildcat was flown by Langley pilots in early 1942. The prototype of the Wildcat series, the XF4F-3, was tested in the Langley Full Scale Wind Tunnel for drag reduction and aerodynamic improvements in 1937. (NASA.)

AVENGER TORPEDO BOMBER. This navy Grumman XTBF-1 Avenger prototype was also flown in early to mid-1942. Flight studies were conducted with the aircraft to verify data and modifications derived from Full Scale Wind Tunnel studies also conducted during 1942. The Avenger was also flown by Langley pilots to develop proficiency in torpedo bomber aircraft. (NASA.)

GULL-WING FIGHTER. The XF4U-1, prototype of the famous navy Corsair fighter series, was evaluated during flight tests at Langley in the early years of World War II. Before the Corsair entered mass production, numerous modifications were made to the design as a result of wind-tunnel tests at NACA Langley, including drag reduction or "cleanup" studies in the Full Scale Wind Tunnel. (NASA.)

MUSTANG PROTOTYPE. The North American XP-51, prototype of the legendary Mustang series, was flight tested at Langley. The aircraft used the revolutionary Langley-developed NACA laminar flow airfoil and was highly regarded and well liked by Langley test pilots. Production versions of the Mustang, deployed to the combat theater, later helped the Allies attain aerial supremacy of the skies over Europe during the war. This aircraft, which has been restored, is now on display at the Experimental Aircraft Association (EAA) Museum in Oshkosh, Wisconsin. (NASA.)

LIGHTNING PROTOTYPE. The second Lockheed P-38 Lightning prototype, known as the YP-38, is pictured during flight tests at Langley. The aircraft underwent drag reduction or "cleanup" studies in the Langley Full Scale Wind Tunnel, and Langley pilots flew the airplane to verify and validate the modifications. It was in the YP-38 that pilots began to notice compressibility problems at high speeds, leading to the initiation of an extensive transonic flight research program at Langley during the latter years of the war. (NASA.)

BREWSTER BUFFALO. A Langley test pilot puts a Brewster F2A-2 Buffalo through its paces in 1942. Like most American fighter aircraft, the Buffalo was tested in the Full Scale Wind Tunnel for drag reduction and flown in modification verification flights. (NASA via Laurence K. Loftin.)

ALL-MOVING HORIZONTAL STABILIZER. This Curtiss XP-42 was used to flight test the all-moving horizontal stabilizer concept beginning in 1942. The aircraft was also used to test specially designed aerodynamic engine cowls. The all-moving horizontal stabilizer concept later proved to be a key in helping to maintain control in transonic and supersonic flight. Most high-performance military jet fighters have all-moving horizontal stabilizers today. (NASA.)

A LATER HAWK. The Curtiss P-40E Warhawk was used to flight test drag-reduction modifications resulting from wind-tunnel studies in the Langley Full Scale Wind Tunnel in 1942. This version of the Warhawk went on to achieve a distinguished combat career in North Africa and the Pacific. (NASA.)

RAZORBACK. The Republic P-47B Razorback Thunderbolt, an early version of the famous Thunderbolt series, was used by NACA Langley in 1942 for loads and maneuverability testing. The heavily armored Thunderbolt went on to excel as an escort fighter for American heavy bombers and in the ground attack role. (NASA.)

BIRDCAGE. The Chance Vought F4U-1 Corsair was an early version of the Corsair series with a "birdcage" canopy. It was test flown by Langley pilots in 1942. Design problems identified included poor vision from the low pilot's seat and the tail wheel configuration. The seat was subsequently raised, along with the tail wheel. These problems had initially hindered the aircraft's carrier trial attempts for the navy, but following the modifications suggested by the NACA, the aircraft passed the carrier trials. (NASA.)

AERODYNAMICS OF THE AIRACOBRA. Langley test pilot Herb Hoover flies a Bell P-39 Airacobra with landing gear down over Langley in early 1943. Note the survey rakes located on the forward fuselage used to measure propeller wake flow. The Airacobra was the subject of much research in Langley's Full Scale Wind Tunnel, where tests led to modifications that drastically improved the performance of the airplane. (NASA via Laurence K. Loftin.)

MASTER OF THE PACIFIC SKIES. In early 1943, this Grumman F6F-3 Hellcat underwent numerous flight tests at Langley. Most notable were performance and handling tests. These tests led to modifications of the aircraft's lateral controls, incorporated in later versions of the Hellcat with improved overall performance. The stalling characteristics of the F6F-3 were also investigated and documented by NACA Langley. Hellcat pilots were ultimately responsible for more than half of all U.S. Naval aerial victories scored in the Pacific theater. (NASA via Laurence K. Loftin.)

RADICAL RAZORBACK. This Republic XP-47F Razorback Thunderbolt was test flown by Langley in early 1943 to evaluate the applicability of the NACA laminar flow airfoil to the Thunderbolt series. No Thunderbolts with laminar airfoils were ever mass-produced. (NASA.)

SWORD OF THE SAMURAI. A captured Japanese Mitsubishi A6M2 Zero arrived at Langley in early 1943 to be outfitted with instrumentation prior to flight test evaluations by the navy. The Allies test flew this machine extensively, and the secrets of the Zero's superiority were revealed. American pilots adjusted their tactics so as to take advantage of the Zero's vulnerabilities, and aircraft manufacturers refined the designs of the Chance Vought Corsair and Grumman Hellcat, which totally outclassed the Zero during the final years of World War II. (NASA via Laurence K. Loftin.)

NAVY DIVE BOMBER. Pictured is a Curtiss SB2C-1 Helldiver dive bomber used by Langley in 1943 for flight studies of the aircraft's control system and maneuverability characteristics. Later in 1944, modifications were made to the aircraft's elevators and wing-tip slots as a result of suggestions recommended by Langley pilots and engineers. The Helldiver went on to become a more-than-adequate replacement for the Douglas SBD Dauntless during the latter stages of the war in the Pacific. (NASA via Laurence K. Loftin.)

LIBERATOR WORKHORSE. This Consolidated B-24D Liberator four-engine heavy bomber was used in performance and handling evaluations and pilot familiarization studies in mid-1943, as well as model bomb drop tests during the latter years of World War II. (NASA.)

LAMINAR FLOW STUDIES. In 1943, Langley conducted extensive flight tests of a North American P-51B Mustang, which used the revolutionary NACA laminar flow airfoil. In this photograph, a wing glove has been fitted over the wing section and a survey rake added to the trailing edge of the wing to measure wake characteristics of the wing section. The use of the laminar flow airfoil helped make the Mustang the premier air superiority piston-engine fighter of World War II. (NASA.)

KINGCOBRA EVALUATIONS. In late 1943, Langley pilots flew this Bell P-63 Kingcobra in studies aimed at improving the aircraft's performance and handling characteristics. One of the resulting aircraft modifications was the addition of a taller vertical tail for improved stability and control. The aircraft had survey rakes installed on the forward fuselage to measure propeller wake flow. (NASA.)

THE BOSS. NACA Langley chief test pilot and head of the Flight Research Division Mel Gough prepares for a research mission in the Bell P-63 Kingcobra in late 1943. Note the tufts of wool attached to the aircraft used to indicate airflow behavior during the test flights. (NASA via Laurence K. Loftin.)

READY FOR RESEARCH. This is a lineup of flight research subjects at NACA Langley in 1944. Note the Sikorsky HNS-1 Hoverfly helicopter parked in front of the flight hangar at the upper right. The Hoverfly was the world's first practical helicopter, and an army version of the helicopter, designated the YR4-B, saw action in the Pacific during the latter part of World War II in search-and-rescue missions recovering downed American airmen. (NASA.)

A BAND OF BROTHERS. From left to right, Melvin Gough, Herbert Hoover, John P. "Jack" Reeder, Stefan A. Cavallo, and William E. Gray Jr. served as the primary test pilots of Langley's Flight Research Division during World War II. The close-knit team shared many accomplishments. The pilots pose in front of a Republic P-47D Thunderbolt fighter. (NASA via Jack Reeder.)

INVADER. In early 1944, Langley began a flight research project aimed at reducing the aerodynamic drag of the engine nacelles on the A-26B Invader bomber. The flight research project led to design modifications that boosted the attack bomber's performance. The Invader served during the latter years of World War II, the Korean War, and during the Vietnam War, a testament to the effectiveness and durability of the design. (NASA via Laurence K. Loftin.)

BUBBLETOP JUG. Nicknamed the "Jug" by its pilots during World War II, the Republic P-47 Thunderbolt was a fighter/bomber mainstay with the Army Air Force. This version of the Thunderbolt, designated P-47D-28, featured a "bubbletop" canopy for improved pilot vision. The airplane was driven by a three-blade prop configuration and was flown in mid-1944 in studies to determine propeller efficiency, handling, and performance studies. (NASA.)

HIGH-ALTITUDE SPITFIRE. A high-altitude British Supermarine Spitfire Mk. VII fighter/interceptor was flown by Langley pilots in late 1944 in studies designed to alleviate drag problems caused by protrusions from canvas coverings on the outer gun ports on the wings. This research was similar to the studies conducted on the early versions of the Spitfire and Hurricane that had made visits to Langley during the early 1940s. Handling and performance studies and pilot familiarization studies were also conducted with this airplane. (NASA.)

MOSQUITO EVALUATIONS. This British-built DeHavilland F-8 Mosquito all-wooden bomber was flown in stability, control, and pilot familiarization evaluations in late 1944. The swift Mosquito was a thorn in the side of German air defense planners during World War II. (NASA.)

DITCHING TESTS. In 1944, the NACA and Army Air Force jointly conducted a flight test project in which an army Consolidated B-24D Liberator was flown from Langley Field and purposely ditched in the James River. The test provided a wealth of data for Allied aircrews and aircraft designers regarding how to best survive a water impact in a large four-engine heavy bomber. In the background can be seen the James River Bridge, which served as a route to and from Southside Hampton Roads. (NASA.)

DAMAGE SURVEYED. A giant crane on a barge hoists the B-24D from the waters of the James River after the ditching test for damage inspection. (NASA.)

"SLOW BUT DEADLY." Douglas SBD-5 Dauntless dive bomber, nicknamed "Slow But Deadly" by its aircrews, was test flown by Langley in late 1944 in handling and performance and pilot familiarization studies. The Dauntless became a force for the navy in the Pacific, noted for the destruction of Japanese warships at the Battle of Midway. SBD-5s were also flown by the Marine Corps, providing air support for ground forces during the "Island Hopping" campaign. (NASA.)

THUNDER IN THE AIR. Langley test pilot Jack Reeder pilots a Republic P-47D-30 Thunderbolt over the James River in early 1945. Langley used its P-47D-30 to study propeller characteristics for several propeller blade configurations. The aircraft featured two propeller wake survey rakes on the forward fuselage used to measure propeller wake distribution. The airplane pictured possessed three propeller blades, in contrast to the four blades that were common on most Thunderbolts that saw combat during World War II. (NASA.)

ADVANCED MUSTANG. In early 1945, Langley test pilots flew this North American P-51D Mustang in handling and performance studies as well as evaluations of the airplane's bubbletop canopy, which permitted better all-around vision for the pilot compared to the framed canopies on earlier versions of the Mustang. The "D" model Mustang was the most widely produced version of the aircraft series. (NASA via Laurence K. Loftin.)

NACA/NASA TEST PILOT LEGEND. John P. "Jack" Reeder, seen here boarding a P-51D Mustang in 1945, initially went to work at Langley in 1938 as an aeronautical engineer working in the Full Scale Wind Tunnel. A shortage of test pilots enabled him to transfer to the Flight Research Division in 1942. Reeder had a 42-year career with NACA/NASA, retiring in 1980, during which he test flew 235 different types of airplanes and rotorcraft. He is considered by many to have been the agency's best test pilot. (NASA via Laurence K. Loftin.)

BEARCAT STUDIES. Langley test pilot William Gray Jr. performs a test flight in Grumman's exceptional XF8F-1 Bearcat prototype in early 1945. The Bearcat rolled off the production line in early 1945 but came too late to see any combat during the war. Langley flew the Bearcat in stability and control studies, which resulted in a production modification to the airplane's vertical tail. The tail was made taller than in the original prototype, resulting in an improvement of the airplane's directional stability and control characteristics. (NASA via Laurence K. Loftin.)

HOVERING HOVERFLY. Jack Reeder maintains a hover in the Sikorsky HNS-1 Hoverfly in March 1945. The Hoverfly was this nation's first practical helicopter, and the NACA made groundbreaking flight studies in the rotorcraft at Langley. Reeder, the agency's first helicopter test pilot, produced the NACA's first formal documentation on handling qualities for practical helicopters resulting from his test flights in the HNS-1. The crosses visible on the fuselage of the helicopter were visual recognition aids. (NASA.)

CONSOLIDATED CAT. The Consolidated PBY-5A Catalina flying boat, shown at NACA Langley in 1945, was used in flight and operational studies aimed at verifying the effects of modifications to the aircraft's hull and spray characteristics brought about as a result of tests on models of the flying boat in Langley's Tow Tank No. 1 facility. The studies helped to alleviate and solve problems with spray washing onto the windshield and engine components during routine sea operations. The Catalina became one of the American legends of World War II, performing crucial search-and-rescue, reconnaissance and maritime patrol, and anti-submarine missions for the navy. (NASA via Laurence K. Loftin.)

WING FLOW MUSTANG TEST BED. Jack Reeder performs a research flight in a wing-flow test North American P-51D Mustang in 1945. Fuselage halves of models with various types of wings were installed on the wing test section of the P-51, which was flown in a steep dive to attain data at near transonic speeds. This transonic research method provided aircraft designers with valuable information before the sound barrier was penetrated in October 1947. (NASA via Jack Reeder.)

DELTA WING TEST MODEL ON P-51D WING TEST SECTION. A delta wing semispan model is mounted on the wing test section of a wing flow P-51D Mustang test bed. A wide variety of wing types were tested near transonic speeds. (NASA via Laurence K. Loftin.)

B-17G MOTHER SHIP. This Boeing B-17G Flying Fortress heavy bomber was used as a mother ship to drop "free-fall" bomb models for transonic research studies in 1945. (NASA.)

DROP MODEL MOUNTED ON UNDERWING PYLON OF B-17G MOTHER SHIP. A free-fall bomb drop model is mounted on the underwing pylon of the B-17G mother ship at NACA Langley in 1945. As the bomb drop model fell through the transonic speed region, it was tracked via telemetering. The data generated from the studies proved to be invaluable to scientists and engineers studying aerodynamics at transonic speeds. (NASA via Laurence K. Loftin.)

BOEING B-29 BOMB DROP MODEL TEST BED. This Boeing B-29 Superfortress heavy bomber was used by NACA Langley in 1945 as a bomb drop model mother ship. (NASA.)

FREE-FALL BOMB DROP MODEL MOUNTED ON UNDERWING PYLON. A close-up view shows a free-fall bomb drop model mounted on the underwing pylon of a Boeing B-29 mother ship. (NASA via Laurence K. Loftin.)

Four

THE JET AGE
1946–1957

During World War II, several nations developed jet fighters. Germany had its swept-wing Messerschmitt Me-262 Schwalbe (Swallow), the world's first operational combat jet fighter, and America had the relatively unsuccessful Bell P-59 Airacomet and the more successful Lockheed P-80 Shooting Star. During this period, it became clear that jet aircraft were the way of the future and that the days of the piston-engine fighter were rapidly coming to an end.

Meanwhile, NACA and Air Force officials began planning the development of a new supersonic research aircraft, the Bell XS-1. The aircraft, dubbed "the flying bullet," was developed by Bell Aircraft; aerodynamic testing of the design in wind tunnels and the NACA portion of the flight test program was managed by Langley engineer John Stack. The Air Force used one of the XS-1 aircraft to conduct its own flight test program, in which Capt. Chuck Yeager served as the main test pilot. Yeager became the first pilot to successfully penetrate the sound barrier. A total of 13 Langley engineers and test pilots were detached to Muroc, California, to conduct the NACA portion of the XS-1 flight tests. Two Langley pilots flew the XS-1 past the sound barrier: Herbert Hoover, the second person and first civilian to break the sound barrier, and Robert Champine. While at Muroc, Champine also made subsonic flights in the Douglas D-558 Phase I Skystreak and Douglas D-558 Phase II Skyrocket.

Back at Langley, aerodynamicists began to explore the benefits of swept wings predicted by Langley engineer Robert T. Jones and German aerodynamicist Adolf Buseman during World War II. Swept wings were later incorporated on two of America's most successful military aircraft of the 1950s, the North American F-86 Sabre jet fighter and the Boeing B-47 Stratojet bomber.

NACA Langley assisted the military in some of the first radar target acquisition and tracking studies for fighter aircraft. Engineers and flight personnel at Langley developed the first fly-by-wire control system for jet fighters, now a standard control feature on modern jet fighters.

The 1950s saw a dramatic increase in the number of flight research projects undertaken by the NACA at Langley. In one year during the 1950s, there were a total of 54 aircraft being tested.

THE "FLYING BULLET" AND ITS MOTHER SHIP. The XS-1-2, flown past the sound barrier by NACA Langley test pilots Herbert Hoover and Robert Champine in 1948, and its B-29 mother ship sit on the tarmac at Muroc, California. (NASA.)

MACH MASTERS. NACA Langley test pilots Robert Champine (left) and Herbert Hoover (right) are seen in front of the XS-1-2 at Muroc, California. Champine performed 13 flights in the XS-1, while Hoover successfully completed 14 flights in the aircraft. Both men made supersonic flights in the XS-1, with Hoover becoming the second person and first civilian to break the sound barrier in 1948. (NASA.)

LANGLEY TEST PILOT ROBERT CHAMPINE AND DOUGLAS D-558-I SKYSTREAK. Langley test pilot Robert Champine performed a total of eight subsonic test flights for the NACA in the Douglas D-558-I Skystreak at Muroc, California, beginning in 1949. These flights were designed to generate data regarding the handling qualities of the aircraft. (NASA via Robert Champine.)

DOUGLAS D-558-II SKYROCKET. The NACA Douglas D-558-II Skyrocket is pictured at Muroc, California. Robert Champine performed the initial test flights of this aircraft for the NACA in 1949. During one flight, he attained a speed of Mach 0.87 in a dive. The flight generated valuable data on the pitch-up problems of high-speed swept-wing aircraft. In 1953, NACA test pilot Scott Crossfield attained a speed of Mach 2 in the aircraft, which was outfitted with a rocket engine. (NASA via Robert Champine.)

SWEPT-WING TEST BED. This unique swept-wing Bell L-39 was test flown by Langley pilots in 1946 in studies aimed at investigating flying qualities of sweptback wings at low speeds. The aircraft was a modified Bell P-63 Kingcobra outfitted with swept wings and used extensively by the navy in their initial swept-wing test program. (NASA.)

BLACK WIDOW. Robert Champine poses next to a Northrop P-61 Black Widow at Langley Field in 1946. NACA Langley flight tested this aircraft in some of its first studies on radar target acquisition and tracking. During the studies, Langley used its vast fleet of P-51 Mustangs as targets for the P-61. Later, when radar was used in jet aircraft, the NACA P-51s were again utilized as target aircraft to test and study the radar systems. (NASA via Robert Champine.)

FIRST JET FIGHTER RESEARCH AIRCRAFT. The Lockheed P-80 Shooting Star served as Langley's first jet fighter flight test subject in 1946. This aircraft was used for pilot familiarization, handling, and performance studies. (NASA.)

TWIN MUSTANG. The North American XP-82 Twin Mustang was used by Langley during the latter 1940s and early 1950s, primarily as a test bed for launching various rocket and missile models, in addition to dropping bomb models. Here the aircraft is seen preparing for a research mission to test the 16-C-1 ram-jet missile, mounted under the wing. The Twin Mustang, later designated F-82, was used extensively during the Korean War both as a night fighter and escort fighter for heavy bombers. (NASA.)

FIRST JET BOMBER FLIGHT RESEARCH SUBJECT. An Air Force North American B-45A, America's first jet bomber, was flown by Langley during the latter 1940s and early 1950s in a variety of flight studies, including pilot familiarization, performance, and handling tests. Tragically, Langley test pilot Herb Hoover lost his life in this particular aircraft when it experienced an in-flight catastrophic structural loads failure in 1952. (NASA.)

AMERICA'S FIRST TAILLESS SWEPT-WING JET FIGHTER DESIGN. A navy Vought XF7U-1 Cutlass tailless swept-wing jet fighter was test flown by Langley in 1948. The aircraft was flown in pilot familiarization, performance, and handling studies. (NASA via Laurence K. Loftin.)

BENDIX MODEL K "WHIRLAWAY" COAXIAL HELICOPTER. Jack Reeder shows off the capabilities of the Bendix Model K "Whirlaway" to members of the military services at Langley Field in 1948. The handling and performance qualities of the coaxial helicopter were extensively tested at Langley. (NASA via Jack Reeder.)

KOREAN WAR LEGEND. The Bell 47 helicopter, which became legendary with the army during the Korean War airlifting wounded American soldiers from the combat front to army MASH (Mobile Army Surgical Hospital) units, was test flown at Langley during the latter 1940s in flying quality studies. (NASA via Frederic Gustafson.)

THUNDERJET. In 1949, Langley pilots flew this Republic YF-84A Thunderjet prototype in performance and handling studies. The Thunderjet saw extensive combat duty during the Korean War with the Air Force. (NASA via Laurence K. Loftin.)

WORLD WAR II HOLDOVER. Later versions of the Chance Vought F4U Corsair, such as this F4U-4B seen here at Langley in 1950, served once again with the U.S. Navy and Marine Corps during the early portion of the Korean War, a testament to the design of the aircraft. This NACA F4U-4B was flown in performance and handling studies at Langley. (NASA.)

THE WEST AREA FLIGHT HANGAR, APRIL 1950. Shown is the construction progress of the West Area Flight Hangar (Building 1244) in April 1950. The aircraft hangar was completed later that same year and was larger than two football fields. It was capable of housing two Convair B-36 Peacemaker bombers, the largest aircraft in the U.S. military arsenal at the time. The two NACA hangars in the East Area were no longer used by the agency following the completion of the West Area Hangar. (NASA.)

PREPARING FOR A FLIGHT RESEARCH MISSION. Pilot Herb Hoover, outfitted with the latest flight gear, enters the cockpit of a NACA F-51 Mustang before taking off on a flight research mission in 1951. Tragically, Hoover lost his life on August 14, 1952, when his B-45A Tornado jet bomber experienced an in-flight structural failure. (NASA via Laurence K. Loftin.)

FLYING A HELICOPTER BLIND. Langley test pilots Jack Reeder and James Whitten (concealed by a drawn shower curtain in the rear portion of the cockpit) fly a NACA H03S-1 helicopter in blind flying experiments in 1951. Whitten, seated in the rear seat, piloted the rotorcraft while Reeder served as the safety pilot seated in the front. The experiments were aimed at evaluating instrument flying for helicopters. (NASA.)

THE "FLYING BANANA." James Whitten (left) and Jack Reeder are shown in conversation following an evaluation of the Piasecki HRP-1 tandem rotor helicopter, dubbed the "Flying Banana" by aircrews in the U.S. Navy and Marine Corps. The HRP-1 was the world's first practical tandem-rotor helicopter, and the NACA conducted handling and stability evaluations of the rotorcraft in 1951. (NASA.)

COLD WAR WARRIOR. Langley used this North American F-86A Sabre Jet, obtained from the Air Force in 1951, in evaluations of the handling and flight characteristics of the swept-wing jet fighter, as well as the flutter characteristics of a swept-wing jet fighter design at high speeds. These studies led to a marked improvement in the maneuverability of the aircraft, which outclassed the Russian-built MIG-15 in the skies over Korea during the Korean War. (NASA.)

NAVY BANSHEE. One of the navy's first jet aircraft, a McDonnell F2H-1 Banshee, undergoes a moment of inertia determination in the Langley West Area hangar in 1951. (NASA.)

SKYKNIGHT STUDIES. During the early 1950s, Langley conducted performance, handling, radar target tracking and acquisition, and pilot familiarization studies in the Douglas F3D-2 Skyknight, the U.S. Navy and Marine Corps' first jet night fighter and attack aircraft. The Skyknight saw service in both the Korean and Vietnam conflicts. (NASA.)

AMERICA'S FIRST SWEPT-WING JET BOMBER. In 1952, Langley pilots flew the Boeing B-47A Stratojet in handling, control, and performance studies. The Stratojet, which served with the Air Force's Strategic Air Command (SAC) and was capable of delivering a nuclear payload to a target, is seen here taking off with the use of a rocket-assisted takeoff (RATO) in addition to normal jet-engine power. (NASA via Laurence K. Loftin.)

SUPERSONIC PROPELLER TEST BED. This McDonnell XF-88B supersonic propeller test bed was used by Langley in 1953 to test the practicality of supersonic propellers on high-speed aircraft. The NACA tested a wide variety of propeller blades, blade configurations, and spinners during the test program. While the program was eventually terminated because of waning interest in high-speed propellers, the Langley XF-88B studies contributed to the development of turboprop engines ultimately used on airplanes including the Lockheed Electra and navy Lockheed P-3 Orion. (NASA.)

A CROWDED HOUSE. A plethora of flight research subjects crams the confines of the Langley Flight Research Hangar during Hurricane Hazel, which hit the Hampton Roads area hard in 1954, causing extensive flooding. The multitude of NACA aircraft in the hangar was typical of the amount of flight research work allotted for the laboratory at the time. (NASA.)

SCORPION. One of the numerous flight test projects undertaken in 1954 was test flying and assessing the radar tracking capability of one of the Air Force's first all-weather jet interceptors, the Northrop XF-89 Scorpion. In addition to being equipped with a radar system, the aircraft was the first in the American arsenal to carry a nuclear air-to-air missile, the Genie. (NASA.)

FLY-BY-WIRE TEST BED. This Grumman F9F-2 Panther was outfitted with an experimental analog fly-by-wire aircraft control system, the precursor to today's highly sophisticated digital fly-by-wire aircraft control systems. The aircraft was flown extensively by Langley in 1954 to test the practicality of the automated control system, which allowed the pilot to control the aircraft by means of a side-stick controller. Fly-by-wire control systems are now standard features on most high-performance military and civil aircraft. (NASA.)

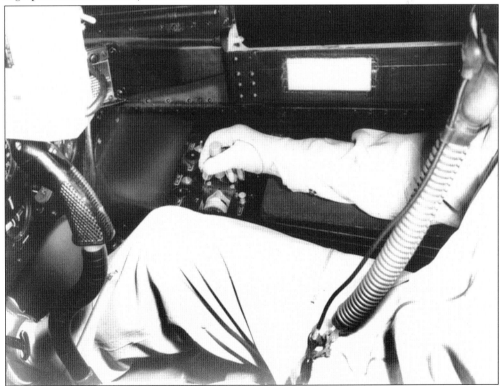

THE AUTOMATED AIRCRAFT CONTROL SYSTEM. NACA Langley test pilot William L. Alford shows off the automated aircraft control system in a publicity photograph. Unlike the high-tech state-of-the-art digital fly-by-wire systems of today, the research aircraft control system operated by means of an analog computer system. (NASA.)

NAVY COUGAR. In 1954, Langley pilots test flew this Grumman F9F-7 Cougar, essentially a Panther with swept wings, in carrier approach, piloting instrumentation, and auto throttle system studies. (NASA via Jack Reeder.)

GUST ALLEVIATION TEST BED. In 1955, Langley researchers used this modified navy Beech C-45 twin-engine light transport to test a specially designed system to help alleviate the effects of wind gusts on aircraft in flight. The system, designed by Langley engineers, was successfully demonstrated and consisted of a special gust sensor in the nose of the aircraft and a hydraulic system that regulated inputs to the wing flaps. (NASA.)

NAVY CRUSADER. In 1956, Langley acquired a navy Vought F8U-1 Crusader for extensive flight testing. The Crusader proved to be an outstanding combat aircraft in the Vietnam War. However, the aircraft initially possessed several nagging mechanical and stability and control problems. A joint research program between Chance Vought, the navy, and the NACA included both wind-tunnel and flight testing. As a result of this research, the aircraft's control system was modified and the problems were cured. (NASA.)

SUPERSONIC VOODOO. Pilot Jack Reeder lands the NACA F-101A Voodoo after performing a research mission in 1956. Langley used the Voodoo for sonic boom studies. Data generated from the studies aided planning for the nation's supersonic transport development efforts in the 1960s. (NASA via Jack Reeder.)

THREE ERAS OF AMERICAN FIGHTER DEVELOPMENT. Three eras of American fighter development are visible inside the Langley hangar in 1956: (from top to bottom) North American F-86A Sabre (1950s), McDonnell F-101A Voodoo (late 1950s to 1960s), and North American P-51D Mustang (1940s). (NASA via Laurence K. Loftin.)

NAVY TIGER. In 1957, Langley conducted flight tests with the first navy jet fighter to incorporate Langley engineer Richard Whitcomb's area-rule concept, the Grumman F11F-1 Tiger. Langley flew the aircraft in tests designed to correlate data generated from wind-tunnel studies concerning area rule. (NASA.)

SUPER SABRE. In 1957, Langley used this North American F-100C Super Sabre to supplement the sonic boom studies that had been carried out earlier with the McDonnell F-101A Voodoo. In addition, research on the low-speed landing approach qualities of a swept-wing supersonic jet was conducted. Langley's F-100C Super Sabre flight studies also led to modifications in the aircraft's design, including a taller vertical tail and improved wing sections. (NASA.)

LANGLEY TEST PILOT WILLIAM "BILL" ALFORD. Pilot Bill Alford is strapped inside the F-100C Super Sabre. Tragically, he perished in an accident that occurred while testing a prototype Royal Navy Blackburn NA.39 Buccaneer strike fighter in England on October 12, 1959. (NASA via Jack Reeder.)

LANGLEY CHIEF TEST PILOT AND HEAD OF FLIGHT OPERATIONS JACK REEDER. Jack Reeder enters the cockpit of the F-100C Super Sabre before taking off on a test flight. (NASA via Laurence K. Loftin.)

Five

THE NASA ERA BEGINS
1958–1970

When the Soviet Union successfully placed Sputnik, the world's first satellite, in Earth orbit in 1957, American military, political, and aerospace leaders realized the urgency of the situation. An advanced phase of the cold war had begun, the Space Race. To match and surpass Soviet achievements in space, President Eisenhower authorized the incorporation of the National Advisory Committee for Aeronautics (NACA) into a new agency known as the National Aeronautics and Space Administration (NASA) on July 29, 1958. Formal NASA operations commenced on October 1, 1958, and the NACA Langley Aeronautical Laboratory (LAL) officially became known as the NASA Langley Research Center (LaRC).

The mission of the new agency was to continue to pursue important aeronautics research but also to conduct space exploration research. Drawing on the broad experience and expertise in the area of flight simulation and training at Langley, NASA headquarters authorized a directive that called for Langley pilots to train the men with the "Right Stuff," the nation's first astronauts. Throughout the 1960s, Langley researchers and pilots developed and assessed new space flight and moon landing simulators that played a critical role in preparing the astronauts for their space missions. When the astronauts came to Langley to use the simulators, they underwent hours of instruction and preparation with Langley's test pilots.

The creation of NASA also brought about drastic mission directive changes at many of the agency's research centers. One of the most dramatic changes was in the area of flight research at Langley, when programs involving high-performance fixed-wing aircraft were transferred to the NASA Flight Research Center at Edwards Air Force Base (AFB), California, and Langley's remaining flight research was cut back to helicopter and V/STOL (Vertical/Short Takeoff and Landing) operations.

A program using unpowered drop models to study spin entry for military aircraft configurations was initiated using Langley support helicopters. The advent of large commercial jet airliners and their routine operation brought about concern from the public regarding jet noise. As a result, Langley undertook an extensive test program during the mid-1960s using the center's first large "flying laboratory," the Boeing 367-80B (B-707 prototype), to study ways of preventing excessive jet noise. In addition, important blown-flap short takeoff and landing (STOL) studies for large jet transports and supersonic transport (SST) in-flight simulation studies were conducted using the B-367-80B.

NASA Sabre Dog. This is a North American F-86D "Sabre Dog" flown by Langley pilots in research studies. A "NASA" designation appears in the old NACA wings background on the tail, denoting that the airplane was in service during the NACA transition to the NASA in October 1958. Following a NASA headquarters directive, all high-performance fixed-wing research aircraft, including this Sabre Dog, were transferred to the NASA Flight Research Center (now known as the NASA Dryden Flight Research Center) at Edwards AFB, California. (NASA.)

"Godspeed John Glenn." Langley test pilot Robert Champine (face visible in middle) and U.S. Marine Corps pilot and astronaut John Glenn are in the centrifuge at the Naval Air Station at Johnsville, Pennsylvania. Glenn was being prepped for tests in the centrifuge as part of his astronaut training. Langley pilots played an instrumental role in helping to train and prepare America's first astronauts for their demanding space missions. (Robert Champine.)

RENDEZVOUS DOCKING SIMULATOR RESEARCH. During the 1960s, Langley Flight Research pilots tested the Rendezvous Docking Simulator, erected along the ceiling of the flight hangar, before the astronauts came to the research center to train on the equipment. Here in a time-lapse photograph, a Gemini space capsule mock-up is seen docking with an Agena Target Booster mock-up during a study using the Rendezvous Docking Simulator. (NASA.)

LUNAR LANDING PRACTICE. Langley test pilot Robert Champine test flies the Lunar Lander Research Vehicle (LLRV) in a simulated lunar landing during the 1960s at the Lunar Landing Research Facility at Langley. Champine's test flights helped to identify and work out problems with the vehicle. (NASA.)

TWIN-ENGINE TILT-WING WONDER. NASA Langley chief test pilot Jack Reeder performs a short vertical takeoff in the Vertol VZ-2 twin-engine tilt-wing V/STOL test bed at Langley in 1960. Langley flight tested the VZ-2 throughout the early 1960s, and the aircraft program proved to be one of the most successful V/STOL programs in the country. The VZ-2 program not only proved the practicality of the tilt-wing V/STOL concept, but it also contributed data and knowledge for more advanced tilt-wing aircraft such as the LTV-Hiller-Ryan XC-142A transport. (NASA.)

TILT-DUCT FAN V/STOL TEST BED. In 1960, Langley and the army conducted flight tests of the Doak VZ-4 tilt-duct fan V/STOL test bed. The VZ-4 flight test program proved the practicality of the tilt-duct concept, which led to further studies by the navy in the Bell X-22A. While the concept held promise for both military and civil applications, it was never used in any production aircraft designs. (NASA.)

BLOWN-FLAP INVESTIGATION. In 1965, a series of flight investigations used the Boeing 367-80B or "Dash 80," prototype of the B-707 jet airliner, to investigate the short takeoff and landing capabilities of the aircraft outfitted with blown flaps to increase lift at low speeds. The airplane became the first large commercial jet transport to serve as a "flying laboratory" at Langley. (NASA.)

JET NOISE-REDUCTION STUDIES. During the mid- to late 1960s, the first NASA noise-reduction studies for large commercial transports were performed with the Dash 80 airplane. (NASA.)

SUPERSONIC TRANSPORT STUDIES. During the mid- to late 1960s, Langley used the Dash 80 to conduct the nation's first supersonic transport (SST) in-flight simulation studies. The research yielded a data and knowledge base that contributed heavily to the first U.S. SST program in the late 1960s and 1970s. (NASA.)

SUPPORT HELICOPTER. Langley's Bell UH-1 support helicopter, with Northrop T-38 Talon unpowered drop model attached, begins to gain altitude over the Plum Tree Flats area near Langley. During the latter 1950s, a program was initiated to study the spin-entry characteristics for military aircraft configurations using dynamically scaled models dropped from Langley support helicopters. The models were controlled by research pilots on the ground. (NASA.)

HOVERING JUMP JET. Test pilot Jack Reeder maintains a hover in the Hawker Siddley P-1127 Harrier prototype. During the early 1960s, Reeder was detached to England to help the British work out problems with the Vertical Short Takeoff and Landing (V/STOL) jet fighter. In addition to wind-tunnel studies that demonstrated the worth of the design, Langley test pilots Jack Reeder and Lee Person, flying the XV-6 Kestrel (a later developmental model of the Harrier), demonstrated the versatility of the basic Harrier design. (NASA.)

KESTREL TEST BED. Pilot Lee Person poses beside the Hawker Siddley XV-6 Kestrel developmental test bed. The Kestrel was flown in Vectoring in Forward Flight (VIFF) studies at NASA Langley during the latter 1960s and early 1970s. The flights demonstrated the feasibility of VIFF and the advantages it afforded a pilot during close-in air combat. (NASA.)

TRAILING VORTEX RESEARCH. Langley used this modified Douglas C-54G Skymaster to study the potential of alleviating dangerous trailing vortices on aircraft during the latter 1960s and early 1970s. The airplane was equipped with trailing-vortex attenuation splines on its wing tips for vortex alleviation. (NASA via Laurence K. Loftin.)

MULTI-ENGINE TILT-WING WONDER. A Ling-Temco-Vought XC-142 tilt-wing Vertical Takeoff and Landing (VTOL) test bed performs a short takeoff at Langley Field in 1970. NASA research pilots flew the XC-142 during the latter 1960s on research missions that proved the viability of the tilt-wing VTOL concept for both civil and military large transport applications. (NASA.)

Six

THE MODERN AGE
1971–2007

By the early 1970s, NASA's space activities had captured the attention of the international media and the world. America was winning the Space Race with the Russians, but the first "A" in NASA, which stands for aeronautics, had not been forgotten. Some of the agency's most important aeronautics work was yet to be accomplished, and flight research, particularly at NASA Langley, would prove to play an instrumental role in helping the nation remain a world leader in aircraft operational safety, atmospheric sciences, and advanced general aviation aircraft development.

During the early 1970s, Langley performed pioneering flight research by perfecting and demonstrating the effectiveness of Microwave Landing Systems for large commercial jetliners. This research was performed with Langley's Boeing 737 Terminal Configured Vehicle (TCV), which was used in Advanced Transport Operating System (ATOPS) research, as well as critical wind shear and artificial vision studies during the latter 1980s and early 1990s. Langley's Convair F-106B Delta Dart was flown in storm hazards (lightning) studies throughout the 1970s and 1980s and in important vortex-flap research during the early 1990s that improved the performance of delta-wing designs.

In response to a sharp increase in the number of general aviation accidents nationwide in the 1970s, Langley initiated a highly successful stall/spin research program for general aviation aircraft. Nearly two decades later, a general aviation technology revival was spurred on by the Langley-led Advanced General Aviation Transport Experiments (AGATE) program which included flight tests of advanced technologies.

Important atmospheric research studies were performed using Langley flight research aircraft. These studies included Earth Bio Mass and Shuttle Exhaust Particle Experiment (SEPEX) Plume Analysis studies to detect dangerous particulates in the atmosphere left by exhaust plumes from the solid rocket boosters of the space shuttle after launch.

During the mid-1990s, Langley acquired the last of its "flying laboratory" aircraft, a Boeing 757, which was used until 2006 in aviation safety and synthetic vision studies.

ATMOSPHERIC TURBULENCE TEST BED. This Martin B-57 Canberra was used by NASA Langley to study characteristics of atmospheric turbulence in 1971. (NASA.)

PREPARING FOR AN ATMOSPHERIC RESEARCH FLIGHT. Langley test pilot Robert Champine waves before boarding the NASA B-57 Canberra for an atmospheric research flight. (Robert Champine.)

STATE-OF-THE-ART SEA KING. In 1972, NASA Langley pilots performed approach and landing studies using this Sikorsky SH-3A Sea King helicopter, which was outfitted with state-of-the-art television and monitor displays. The studies were aimed at improving safety, efficiency, and preciseness of rotorcraft approaches and landings. (NASA.)

HELICOPTER GUNSHIP. This Huey AH-1G Cobra helicopter gunship, acquired in 1972, was flown at Langley throughout the 1970s in tests aimed at investigating the aerodynamics of main rotor blade airfoil sections and acoustic effects. (NASA.)

AUTOMATIC VERTICAL LANDING TEST BED. This Boeing Vertol 107 (CH-46C Sea Knight) was used by Langley researchers during flights at NASA Wallops Island to showcase the first-ever automatic vertical landing system for rotorcraft in 1972. (NASA.)

SKYCRANE HEAVY LIFTER. In 1973, NASA Langley studied the heavy lifting capability of the Sikorsky CH-54B Skycrane as demonstrated in this photograph, in which the helicopter can be seen preparing to lift a T-38 Talon jet trainer. (NASA via Laurence K. Loftin.)

BOEING B-737 TERMINAL CONFIGURED VEHICLE (TCV). In 1973, NASA Langley obtained a Boeing 737 jetliner (the original 737 prototype) for use in its Terminal Configured Vehicle (TCV) Program. Under this program, Langley engineers and pilots effectively demonstrated the new Microwave Landing System (MLS), which enhanced jetliner safety in approaches and landings at airports. The MLS was later selected by the International Civil Aviation Organization (ICAO) to replace the older Instrument Landing System (ILS). (NASA via Laurence K. Loftin.)

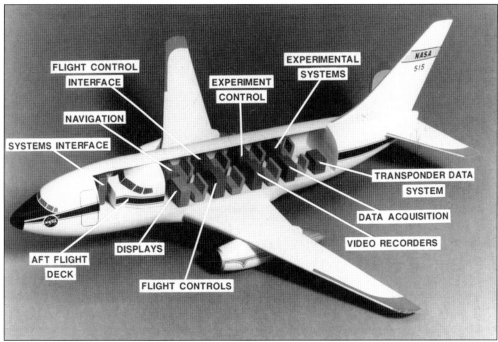

BOEING B-737 TCV CUTAWAY. Langley's TCV program called for the development and flight testing of advanced flight-control and cockpit-display systems that were evaluated by a pilot in an aft flight deck incorporated within the B-737. This cutaway diagram illustrates the aft flight deck and aircraft systems. (NASA.)

Original Boeing B-737 TCV Front Flight Deck. This view shows the maze of antiquated analog displays in the original Boeing B-737 TCV front flight deck before installation of TCV program glass cockpit displays, advanced avionics, instruments, and control systems. (NASA.)

Advanced B-737 TCV Flight Deck. This is the advanced Boeing B-737 TCV flight deck following installation of TCV program glass cockpit displays, advanced avionics, instruments, and control systems. Highly impressed with NASA's pioneering research on glass cockpits in the B-737, visiting Boeing engineers subsequently influenced management to incorporate such systems in future Boeing transports. Glass cockpits are now standard on advanced civil transports and business jets. (NASA.)

GROOVED RUNWAY TESTING.
During the early 1970s, Langley researchers conducted extensive tests of runway grooving at the NASA Wallops Island Flight Facility using a Convair 990 airliner and other aircraft. Runway grooving was designed by Langley researchers to reduce incidences of tire hydroplaning and to enhance braking. The grooving concept is now widely used for improved safety at airports and on highways throughout the United States. Here Langley researchers Tom Yager (left) and Walter Horne survey skid marks on the runway left by the NASA Convair 990 airliner. (NASA.)

VTOL APPROACH AND LANDING TECHNIQUE (VALT) PROGRAM. Beginning in 1974 and lasting throughout the 1970s, NASA Langley conducted the highly successful VALT program aimed at the development and testing of new automated navigational systems for helicopters. Here the VALT test bed, a Boeing Vertol CH-47 Chinook, and VALT personnel pose for a publicity photograph. (NASA via Jack Reeder.)

NASA Langley Flight Research Pilots, Mid-1970s. From left to right are (kneeling) Phil Brown, Perry Deal, Jim Patton, and Fred Gregory; (standing) Bob Champine, Roger Van Gunst, Lee Person, and Dick Yenni. Fred Gregory, who had previously served as a pilot in the U.S. Air Force, later worked as the agency's deputy administrator before retiring. The pilots are posing in front of the Flight Research Hangar (Building 1244) and NASA's B-737 Terminal Configured Vehicle (TCV) used in flight research studies to help pioneer the revolutionary microwave landing system (MLS). (NASA via Robert Champine.)

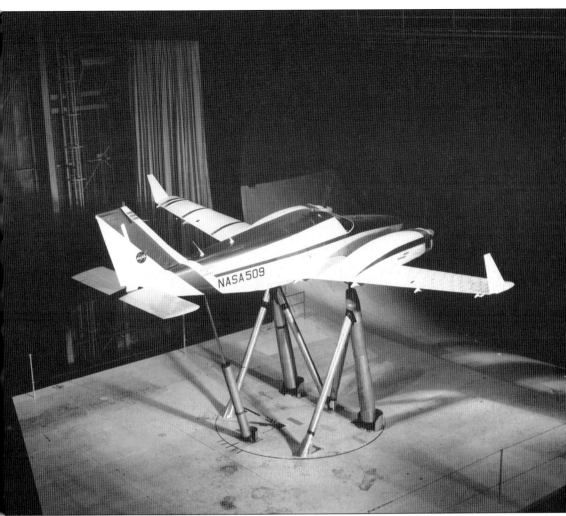

Advanced Technology Light Twin (ATLIT) Test Bed. In 1977, Langley and the University of Kansas conducted flight tests of the Advanced Technology Light Twin-Engine Airplane (ATLIT) to demonstrate and study the performance advantages of several technologies, including an advanced airfoil. The aircraft was also one of the first general aviation aircraft to possess Langley-developed winglets. Here the aircraft is mounted in the Langley Full Scale Wind Tunnel for aerodynamic tests. (NASA.)

AERIAL APPLICATIONS. During the mid-1970s, Langley researchers used an Ayers Thrush Commander agricultural aircraft in studies of the performance, handling qualities, and aerial dispersal characteristics of an agricultural aircraft. In this picture, Langley pilot Philip Brown performs a research flight in the Thrush Commander over a smoke injector array for wake vortex observation. Langley researchers also analyzed this wake using computational methods to assess its effect on chemical dispersion. (NASA.)

STALL/SPIN RECOVERY PROGRAM AIRCRAFT AND SPIN TUNNEL MODELS, 1978. During the early 1970s, Langley investigated spin-recovery techniques for general aviation (GA) aircraft. In concert with Langley Spin Tunnel tests, flights were made for three modified GA aircraft: a Cessna 172 Skyhawk, a Grumman American AA-1 Yankee, and a Piper PA-28 T-Tail Arrow. The models are shown (from left to right in front) along with the research aircraft (from left to right in back). The research aircraft were outfitted with tail-mounted auxiliary spin-recovery parachute systems for emergency recovery from unrecoverable spins. (NASA.)

MODIFIED GRUMMAN AMERICAN AA-1 YANKEE. The spin-recovery parachute system, designed and developed by Langley researchers, was successfully deployed 28 times, preventing spins that could not be stopped using the conventional aircraft aerodynamic controls. The Langley Stall/Spin Recovery Program led to the incorporation of technology in such contemporary state-of-the-art GA aircraft designs as the Cirrus SR-20. (NASA.)

STALL/SPIN-RECOVERY EXPERT. Langley chief test pilot James Patton is in the cockpit of the modified Grumman American AA-1 Yankee. Successful deployments of the spin-recovery parachute were represented by special symbols on the fuselage. Patton received national recognition for his role in the Stall/Spin Recovery Program when he received the Iven C. Kincheloe Award of the Society of Experimental Test Pilots for the most outstanding U.S. test pilot for 1978. (NASA.)

EARLY STORM HAZARD RESEARCH TEST BED. This DeHavilland Twin Otter was used during the mid- to latter 1970s to study the development of crosswind landing gear systems and for early pathfinder storm hazard research that was greatly expanded with an F-106B jet fighter. (NASA via Laurence K. Loftin.)

CANARD REVOLUTION. This Burt Rutan–designed Defiant featured an unorthodox general aviation design and was evaluated by Langley pilots in 1978. The aircraft features included canards, swept wings with winglets, and tractor/pusher propellers. The airplane was designed to be stall-resistant and easily controllable with only one engine operative. (NASA.)

STORM CHASER. This Convair F-106B Delta Dart aircraft was used by Langley to perform storm hazard research and other studies from the 1970s to the early 1990s. During a research flight in 1984, the airplane was struck by lightning 72 times over a 45-minute time span. Detailed data were obtained with special instrumentation that provided designers with data on environmental conditions for lightning strikes, as well as electrical characteristics, such as intensity and dwell time. (NASA.)

A SHOCKING EXPERIENCE. This is a rear-facing view of Langley researcher Bruce Fisher in the backseat of the NASA F-106B during a lightning strike to the airplane. Note the trailing plasma field leaving the aircraft at its wing tips. Data produced from this research was later used by military and civilian pilots in severe weather avoidance procedures and by aircraft manufacturers in the construction of improved aircraft designs that featured enhanced lightning protection. (NASA.)

LANGLEY FLIGHT HANGAR, 1980. In this view of research aircraft in the Langley Flight Hangar in 1980, maintenance is being performed on the F-106B and Boeing B-737 and a variety of other research aircraft. (NASA via Laurence K. Loftin.)

MOHAWK TEST BED. During the early 1980s, Langley and industry partners conducted flight tests of a Grumman OV-1B Mohawk test bed that carried a specially designed laminar flow engine nacelle pod. The prop on the wing, under which the engine pod was slung, was feathered so as to not trigger turbulence in the airstream. The data obtained were used to validate theoretical predictions by Langley researchers and to determine the effects of factors such as noise on laminar flow. (NASA.)

GENERAL AVIATION NATURAL LAMINAR FLOW FLIGHT TESTING. This Cessna Centurion aircraft was used by Langley and Cessna in a joint program to test natural laminar flow airfoils during the 1980s. Sublimating chemicals are being used to visualize the extent of laminar flow obtained in flight. In the photograph, laminar flow is denoted by a whitish coating. Note the wedge-shaped areas denoting the loss of laminar flow because of bug accretion on the wing leading edge. The flight research helped to provide confidence that such airfoils would be successful with today's construction practices. (NASA.)

PIPER PA-28 T-TAIL WING-TIP TURBINE TEST BED. Within a broad range of exploratory flight research with numerous general aviation designs, this Piper PA-28 T-tail, specially equipped with wing-tip turbines, happened to be one of the most unique. This 1985 photograph shows the aircraft in flight with wing-tip turbines in the "off" position. Langley researchers attempted to harness wake vortex energy generated at the wing tips of the airplane to use as an auxiliary power source for the electrical systems on the aircraft. This airplane was also used in the research center's highly successful Stall/Spin Recovery Program during the latter 1970s and early 1980s. (NASA.)

LAMINAR TURBO MENTOR. This Beechcraft T-34C Turbo Mentor was used by Langley in fundamental research to evaluate the ability of various chemicals and pressure-sensitive coatings to visualize areas of natural laminar flow on a special gloved wing test section during the 1980s. (NASA.)

GATES MODEL 28 LEARJET. This Gates Model 28 Learjet was used by Langley in natural laminar flow studies in 1989. Note the natural laminar flow test section on the left wing. The Model 28 was the first U.S. business jet to use Langley-developed winglets in its design. (NASA.)

Northrop T-38 Talon Jet Trainer. Virtually all NASA research pilots and astronauts flew high-performance military jets to sharpen their skills. Langley pilots flew several aircraft, including this T-38 Talon jet trainer during the 1980s, to maintain piloting proficiency. (NASA.)

HEADING OUT ON A RESEARCH FLIGHT. Langley test pilot Harry Verstynen boards a Northrop F-5F used for pilot proficiency flights before heading out on a research flight in 1990. Although Langley did not normally pursue flight research programs involving high-performance military aircraft, its research pilots participated in piloted simulator evaluations of advanced configurations and concepts devised by Langley researchers, and maintaining proficiency was a requisite for the task. (NASA.)

F-106B VORTEX FLAP TEST BED. During the latter 1980s and early 1990s, Langley used its F-106B to perform vortex flap research to improve the aerodynamic performance of delta-wing aircraft at high angles of attack. Subject of a large NASA and industry effort, the investigation also included analytical and wind-tunnel studies. Note the specially-designed vortex flap extension on the leading edge of the wing. The flap concept controlled the characteristics of the vortical flow shed off the leading edge of the swept delta wing during flight. Mounted on the spine of the fuselage is a scanning flow-visualization system, and the white lines on the upper surface of the right wing are pressure belts. The airplane was officially retired from service in 1992 and is now on display in the Virginia Air and Space Center in Hampton, Virginia. (NASA.)

CESSNA 402 TEST BED. The Cessna 402 pictured here was used by Langley as part of an integrated avionics systems program with special emphasis on applications to single-pilot instrument flight rule (IFR) conditions. The airplane was also used to evaluate concepts for ride comfort instrumentation for GA airplanes, in atmospheric studies designed to analyze the Earth's atmospheric properties as influenced by human practices, and to detect dangerous particulates in the atmosphere left by exhaust plumes from the space shuttle solid rocket boosters after launch. (NASA.)

PILOT PROFICIENCY. This Cessna U-3A (310), shown here in 1990, was used by Langley for pilot proficiency and research support missions. (NASA.)

RESEARCH FLEET, 1991. Langley's flight research fleet, including research and support aircraft, is shown in 1991. (NASA.)

DROP-MODEL TESTING. A Langley support helicopter carries an unpowered model of the X-31 research aircraft to high altitude for drop in 1991. Langley provided extensive wind-tunnel and drop-model support to this project, which focused on post-stall maneuverability. Results from Langley wind-tunnel and drop-model tests were used as real-time inputs and guidance by the U.S./German X-31 flight test team at NASA Dryden. (NASA.)

FIGHTING FALCON. The Lockheed Martin F-16A Fighting Falcon was briefly used by Langley to train its force of flight research technicians and pilots in preparation for a possible research program using the F-16XL in 1991. (NASA.)

NORTH AMERICAN ROCKWELL OV-10A BRONCO. This North American Rockwell OV-10A was used by Langley in wake vortex studies. The experiments were conducted by NASA as part of an effort to provide the technology to safely improve the capacity of the nation's air transportation system, and specifically to provide key data in understanding and predicting wake vortex decay, transport characteristics, and the dynamics of encountering wake turbulence. The OV-10A performed several roles, including meteorological measurements platform, wake-decay quantifier, and trajectory quantifier for wake encounters. Extensive research instrumentation systems included multiple air data sensors, video cameras with cockpit displays, aircraft state and control-position measurements, inertial aircraft-position measurements, meteorological measurements, and an on-board personal computer for real-time processing and cockpit display of research data. (NASA.)

SUPPORT HELICOPTER. This Bell Huey UH-1 Iroquois, shown in 1992, was used by Langley in a research support role. (NASA.)

WIND SHEAR STUDIES. A dramatic photograph shows the Langley B-737 in Orlando, Florida, during wind shear studies in 1992. With special instrumentation, the aircraft verified data and successfully demonstrated the usefulness of airborne detection systems to avoid dangerous wind shear conditions. This highly successful research, which had been initiated as a result of national concern over fatal wind shear accidents during the 1980s, helped stimulate the application of in-flight wind shear detection systems used in commercial airliners today. (NASA.)

HIGH-LIFT RESEARCH TEST BED. The wings of Langley's B-737 were highly instrumented for a special high-lift research program. The objective of the research was to provide data for correlation between computational aerodynamic predictions and flight measurements for the landing configuration with leading- and trailing-edge flaps deflected. Note the wool tufts that were placed on the wing upper surface to permit visualization of flow properties. (NASA.)

HIGH-SPEED CIVIL TRANSPORT HIGH-LIFT TEST BED. This Lockheed Martin F-16XL arrived at Langley in 1993 for planned high-speed civil transport (HSCT) high-lift research. The design of the unique F-16XL wing was a cooperative effort of the Langley staff and the General Dynamics Corporation. The new wing was designed to provide the F-16 aircraft with improved supersonic performance while maintaining comparable transonic performance to that provided by the conventional F-16 design. The aircraft, with its cranked wings, was ideally suited for the HSCT research. The primary objective of the flight test project was to verify the performance of high-lift concepts while ensuring compliance with community noise standards. However, plans for the high-lift program were cancelled, and the aircraft was returned to the NASA Dryden Flight Research Center, where it was used in a cooperative Langley-Dryden study of the vortical flow on the wing, known as the Cranked Arrow Wing Aerodynamics Project (CAWAP). Various flight, wind-tunnel, and computational fluid dynamics (CFD) data sets were generated during the CAWAP. (NASA.)

LANGLEY FLIGHT RESEARCH FLEET, JUNE 1994. From left to right are (foreground) Boeing B-737 Flying Laboratory, Lockheed Martin F-16XL, and Boeing B-757 Flying Laboratory; (background) Bell Huey UH-1 helicopter, Northrop T-38 Talon, Beech BE-80 Queen Air, North American Rockwell OV-10 Bronco, Beechcraft U-21A, and Beech T-34C Turbo Mentor. The NASA OV-10 was used for wake vortex studies; the Lockheed Martin F-16XL for wing vortex flow studies; the B-737 for cockpit display, wind shear, runway friction, and wake vortex studies; and the B-757 for advanced synthetic vision studies, runway friction, aviation safety testing, and studies of flight in adverse weather. (NASA.)

FLYING LABORATORY. In 1994, the Langley Research Center acquired a B-757-200 aircraft to replace the aging B-737 Transport Systems Research Vehicle (TSRV). The mission for the B-757 was to continue the three-decade tradition of civil transport technology research begun by the TSRV. This standard 757 aircraft was transformed into an aeronautical research flying laboratory known as Airborne Research Integrated Experiments System (ARIES), shown in its original Eastern Airlines markings while performing a research flight near the research center in 1996. (NASA.)

RUNWAY FRICTION TESTS. Langley's B-757 performs runway friction tests at Kenneth Ingle Sawyer Air Base in Gwinn, Michigan, in February 1999. Langley's staff collaborated with several international partners in this research area, which was a Langley specialty. The B-757 carried on research legacies started with the Langley B-737 in earlier programs. (NASA.)

DROP-MODEL LAUNCH. A Langley research team performs a successful launch of an unpowered navy F-18E/F Super Hornet drop model from a helicopter near Wallops Island in 2000. This critical research enabled engineers to learn more about the spin resistance of the Super Hornet design. Information gathered from the flight results was directly applied in subsequent full-scale aircraft flight tests. (NASA.)

AGATE-INSPIRED AIRCRAFT DESIGNS. The Lancair Columbia 300 (left) and Cirrus SR-20 (right) are state-of-the-art general aviation aircraft designs inspired by the Langley-led Advanced General Aviation Transport Experiments (AGATE) program. AGATE reinvigorated the general aviation industry in the United States and led to the development of several advanced technologies incorporated in the designs of the Columbia 300 and SR-20. Langley later obtained both a Columbia 300 and SR-20 for flight research duties. (NASA.)

ADVANCED SYNTHETIC VISION TESTING. In 2001, Langley and several industry partners used the B-757 Flying Laboratory to perform synthetic vision tests at the Eagle County Regional Airport near Vail, Colorado. The objective of the studies was to demonstrate the capabilities of synthetic vision systems in an operationally challenging environment with imposing terrain features. (NASA.)

ADVANCED SYNTHETIC VISION SYSTEMS. A close-up view shows the advanced synthetic vision research concept in the cockpit of Langley's B-757 while in flight over Vail, Colorado. These advanced synthetic vision systems have enhanced aviation safety and are rapidly evolving as key factors in the future of all-weather flight operations. In 2006, Langley's B-757 was retired to the NASA Dryden Flight Research Center in California. (NASA.)

ABOUT THE AUTHOR. Author Mark A. Chambers (left) and his father, Joseph R. Chambers (right), pose for a photograph in Langley's Full Scale Wind Tunnel at a ceremony in 1995, when the tunnel was officially decommissioned by NASA and turned over to Old Dominion University for operations. In the tunnel is a B-737 model that was tested in a wake vortex study, the last NASA test to be run in the tunnel. Joseph worked at NASA Langley as an aeronautical engineer for 36 years, serving as a division chief before retiring in 1998. The author worked at the time of this picture as a contractor writer/editor in Langley's Office of Public Affairs. (NASA.)

ACROSS AMERICA, PEOPLE ARE DISCOVERING SOMETHING WONDERFUL. *THEIR HERITAGE.*

Arcadia Publishing is the leading local history publisher in the United States. With more than 3,000 titles in print and hundreds of new titles released every year, Arcadia has extensive specialized experience chronicling the history of communities and celebrating America's hidden stories, bringing to life the people, places, and events from the past. To discover the history of other communities across the nation, please visit:

www.arcadiapublishing.com

Customized search tools allow you to find regional history books about the town where you grew up, the cities where your friends and family live, the town where your parents met, or even that retirement spot you've been dreaming about.